THE POCKET WATCH

THE HISTORY AND STORIES SURROUNDING THE FIRST POCKET WATCHES

British Library Cataloguing-in-Publication Data
A catalogue record for this book is available from the
British Library

Contents

A History of Clocks and Watches

Horology (from the Latin, Horologium) is the science of measuring time. Clocks, watches, clockwork, sundials, clepsydras, timers, time recorders, marine chronometers and atomic clocks are all examples of instruments used to measure time. In current usage, horology refers mainly to the study of mechanical time-keeping devices, whilst chronometry more broadly included electronic devices that have largely supplanted mechanical clocks for accuracy and precision in time-keeping. Horology itself has an incredibly long history and there are many museums and several specialised libraries devoted to the subject. Perhaps the most famous is the Royal Greenwich Observatory, also the source of the Prime Meridian (longitude 0° 0' 0"), and the home of the first marine timekeepers accurate enough to determine longitude.

The word 'clock' is derived from the Celtic words clagan and clocca meaning 'bell'. A silent instrument missing such a mechanism has traditionally been known as a timepiece, although today the words have become interchangeable. The clock is one of the oldest human interventions, meeting the need to consistently measure intervals of time shorter than the natural units: the day, the lunar month and the year. The current sexagesimal system of time measurement dates to approximately 2000 BC in Sumer. The Ancient Egyptians divided the day into two twelve-hour periods and used large obelisks to track the movement of the sun. They also developed water clocks, which had also been employed frequently by the Ancient Greeks, who called them 'clepsydrae'. The Shang Dynasty is also believed to have used the outflow water clock around the same time.

The first mechanical clocks, employing the verge escapement mechanism (the mechanism that controls the rate of a clock by advancing the gear train at regular intervals or 'ticks') with a foliot or balance wheel timekeeper (a weighted wheel that rotates back and forth, being returned toward its centre position by a spiral), were invented in Europe at around the start of the fourteenth century. They became the standard timekeeping device until the pendulum clock was invented in 1656. This remained the most accurate timekeeper until the 1930s, when quartz oscillators (where the mechanical resonance of a vibrating crystal is used to create an electrical signal with a very precise frequency) were invented, followed by atomic clocks after World War Two. Although initially limited to laboratories, the development of microelectronics in the 1960s made quartz clocks both compact and cheap to produce, and by the 1980s they became the world's dominant timekeeping technology in both clocks and wristwatches.

The concept of the wristwatch goes back to the production of the very earliest watches in the sixteenth century. Elizabeth I of England received a wristwatch from Robert Dudley in 1571, described as an arm watch. From the beginning, they were almost exclusively worn by women, while men used pocket-watches up until the early twentieth century. This was not just a matter of fashion or prejudice; watches of the time were notoriously prone to fouling from exposure to the elements, and could only reliably be kept safe from harm if carried securely in the pocket. Wristwatches were first worn by military men towards the end of the nineteenth century, when the importance of synchronizing manoeuvres during war without potentially revealing the plan to the enemy through signalling was increasingly recognized. It was clear that using pocket watches while in the heat of battle or while mounted on a horse was impractical, so officers began to strap the watches to their wrist.

The company H. Williamson Ltd., based in Coventry, England, was one of the first to capitalize on this opportunity. During the company's 1916 AGM it was noted that '...the public is buying the practical things of life. Nobody can truthfully contend that the watch is a luxury. It is said that one soldier in every four wears a wristlet watch, and the other three mean to get one as soon as they can.' By the end of the War, almost all enlisted men wore a wristwatch, and after they were demobilized, the fashion soon caught on - the British Horological Journal wrote in 1917 that '...the wristlet watch was little used by the sterner sex before the war, but now is seen on the wrist of nearly every man in uniform and of many men in civilian attire.' Within a decade, sales of wristwatches had outstripped those of pocket watches.

Now that clocks and watches had become 'common objects' there was a massively increased demand on clockmakers for maintenance and repair. Julien Le Roy, a clockmaker of Versailles, invented a face that could be opened to view the inside clockwork – a development which many subsequent artisans copied. He also invented special repeating mechanisms to improve the precision of clocks and supervised over 3,500 watches. The more complicated the device however, the more often it needed repairing. Today, since almost all clocks are now factory-made, most modern clockmakers only repair clocks. They are frequently employed by jewellers, antique shops or places devoted strictly to repairing clocks and watches.

The clockmakers of the present must be able to read blueprints and instructions for numerous types of clocks and time pieces that vary from antique clocks to modern time pieces in order to fix and make clocks or watches. The trade requires fine motor coordination as clockmakers must frequently work on devices with small gears and fine machinery, as well as an appreciation for the original art form. As is evident from this very short history of clocks and watches, over the centuries the items themselves have changed – almost out of recognition, but the importance of time-keeping has not. It is an area which provides a constant source of fascination and scientific discovery, still very much evolving today. We hope the reader enjoys this book.

t

FAMOUS EARLY POCKET WATCHES

Guido Fawkes's Watch—James I.'s Watches—Sir William Howard's Watch—Prince Henry's Watches—David Ramsey—James I. sends a Watch to his Sons—Small Watch like a Tulip—The Jesuit and his Watch—Milton's Watch—Watch Seals—Clockmakers' Company—Hollar's Representations of Watches—Watch found at Lord Torphichen's Seat—Portrait Watches—Charles I.'s Watches—Henrietta Maria's Gift of a Watch—Van Tromp's Watch—Oliver Cromwell's Watches—Charles II.'s Watchas—Pepys and Lord Brounker's Watch—Watch Key given by Charles II.—Enamelled Watches—Watch Enamellers—Jacinth Watch—Richly Ornamented Watches—Durability of Watches—Novel Trial about a Watch—Advertisements about Lost Watches—Dr. Taswell's Watch—Watch found at Chislehurst—Nautilus-shaped Watch—Invention of the Spiral Spring—Dr. Hooke—Thomas Tompion—Daniel Quare—Edward Barlow—Repeating-Watches—James II.'s Watch—The Anchor Escapement—Touch Watches for the Blind—Law relating to Watches—Simon de Charmes.

In the third year of the reign of James I. a watch was found upon Guido Fawkes, which he and Percy had bought the day before, "to try conclusions for the long and short burning of the touchwood which he had prepared to give fire to the train of powder."

A watch that once belonged to James I., and was formerly in the collection of Sir Ashton Lever, and was purchased at the sale of his museum in 1806, was exhibited and described by Mr. H. Syer Cumming at a meeting of the Archæological Institute, on May 9th, 1855. It is of an oval shape, and measures, without its case, one inch and seven-eighths in length, one inch and three-eighths in width, and five-eighths of an inch in thickness. It is of French manufacture, and is nearly all of brass. The dial is a silver ring one-eighth of an inch wide. The hours are engraved in Roman numerals, with a little stud at the base of each, by means of which the time could be ascertained in the dark by the touch. The steel hand is in the form of a dart. Above the dial is engraved a figure of Leda and the Swan; and below, a cherub's head, on each side of which is a fox; and the space between the devices is filled with rich foliage in scrolls. Within the circle of the dial is a view of a river, with swans, a bridge, and beyond it horses, trees, fields, hills, and a stag-hunt. On the plate at the back is engraved, "R. Dieu, à Paris." The cock which covers the balance-wheel is of rich design. Instead of a chain, this watch has a line of catgut; it has no hair-spring; neither are any screws used in its construction, which circumstances prove its early date.

There is another curious watch of this period still extant. It has a silver case, richly ornamented with subjects from mythology, beautifully chased. It bears the following inscription on the inner rim:—"From Alethea, Covntess of Arvndel, for her deare son, Sir William Howard, K.B., 1629." It is of oval form, two inches and a-half in diameter, and one inch and a-half in thickness. It strikes the hours, has an alarum, shows the days of the week, the age and phases of the moon, the days and months of the year, and the zodiac. On the inside is a Roman Catholic calendar, with the date 1613, and the maker's name, "P. Combret, à Lyons." In the South Kensington Museum is a silver watch shaped like a pecten shell; the dial being chased and engraved with scroll-work. The maker's name is "Pierre Combret à Lyon." The length is two inches and one-eighth, and the width one inch and three-quarters. It was purchased for 8*l.* 10*s.*

At a meeting of the Archæological Institute, held on December 7th, 1849, Mr. Frederick Ouvry exhibited a curious watch, which was supposed to have belonged to James I., or to have been a present from him. It was in shape like an egg flattened. It had an outer case of plain silver. The inner case was beautifully engraved, on one side representing Christ healing a cripple, with the motto used by King James, "Beati pacifici," and the royal arms underneath. On the other side, the Good Samaritan, with the inscription, "S. Lucas, c. 10." Inside the lid was a well-executed engraving of James I., with his style and title. Round the rim were the rose, the harp, and the thistle, all crowned with the initials "J. R." The face had a calendar, and showed the moon's age and other things. On the works was the maker's name, "David Ramsay Scotus me fecit." Underneath a small shield, which concealed the hole for winding, was the name of the engraver, "Gerhart de Heck sculps." It was the property of Miss Boulby, of Durham. It had long been in her family, and was supposed to have come to them from the Russells, of Woburn. The artist Heck mentioned above is not named by Walpole, nor in Bryan's 'Dictionary.' He was possibly of the same family as Nicholas Vander Heck, a painter at the Hague, about 1600, or John Vanden Hecke, who was settled at Antwerp, about 1650. One of the watches which had belonged to James I., was sold at the Duke of Sussex's sale.

In "The Accompte of the Money Expended by Sir David Murray, K^t, as Keaper of the Privie Purse to the late Noble Prynce Henry, Prynce of Wales, from the first of October, 1610, to the sixth of November, 1612 (the daye of the decease of the said Prynce) as lykewise for certaine paymentes made after the deathe of the said Prynce in the monethes of November and December, 1612," in the Audit Office, Somerset House, is the following entry:—"Watches three bought of Mr. Ramsay the Clockmaker, lxj li." (61*l.*)

In the list of 'Guyftes and Rewardes,' in the same account we find to "Mr. Ramsey the Clockemaker, xj s." The money of the reigns of Elizabeth and James is valued at five times its present value.

The Ramsey named above was David or Davie Ramsey, the friend of Master Heriot, Gingling Geordie, and an eminent artificer in the time of James I. and Charles I. He was the first Master of the Clockmakers' Company on their incorporation in the seventh year of the reign of Charles I., 1631. An engraved portrait of George Heriot, jeweller to King James, who died in 1623, aged sixty-three, was published in 1743; the painter being Scougal, and the engraver Joc. Esplens.

On April 18th, 1623, James I. wrote from Windsor a letter to his "sweete boyes," who were then in Spain; and therewith sent certain jewels for their companion, the Duke of Buckingham, to give away. One jewel was, according to a contemporary inventory, "a clocke of goulde, garnisht on the one side with letters of dyamondes: Dieu et mon droyte; and on the other side, a cross of dyamondes fullie garnisht, with a pendante of dyamondes." The king in his letter said "the watche is the richest and fittest for some olde ladie in my opinion." With this royal opinion before us, we are induced to consider whether his majesty thought that a timekeeper was better suited to an elderly lady because of her nearness to the grave; or that the favour and influence of an aged woman, as against those of a young one, were more worth a costly trinket; or that youthful beauty did not require a blaze of diamonds to enhance its attractions. Later we find a warrant of indemnity to the Lord High Treasurer for having delivered certain jewels to James I. that were sent into Spain as above mentioned, dated July 7th, 1623; in which document the "clocke of gould" is named.

In the 'Mirror' for 1836 is an engraving of a pretty, little, ancient watch, which was of silver, and about the size of a large walnut, and might be closed by the fingers in the palm of the hand without being seen. It was shaped like a tulip, the outer case being divided into three equal compartments, ornamented with a sort of leaf on a roughly-chased groundwork, and looking as unlike a watch as it possibly could. These leaves opened a little at the bottom of the watch, and disclosed a small spring, which, on being pressed, pushed up the lid and displayed the dial-plate, contained within a circular border; the space between which and the outer oval boundary of the face was filled by ornamental flower-work. In the centre of the dial-plate, within the figures, the rose and thistle were entwined, which would seem to fix the date of the construction of the watch to the reign of James I. Watches were worn at this period hung round the neck by a chain, and the silver one used for this purpose was still remaining attached to the above article.

In an engraved portrait of Marcus Antonius de Dominis, Archiepiscopus Spalatensis, aged fifty-seven in 1617, appears a watch of curious pattern and antique construction, placed on the table before him, with the lid open, showing the dial. This prelate came into England in the reign of James I., where he published his book 'De Republicâ Ecclesiasticâ,' a portion of which he is represented in the act of writing in the engraving.

Gee, in his 'Foot out of the Snare,' published in 1624, says:—"If about Bloomesbury or Holborne thou meet a good smug fellow in a gold-laced suit, a cloak lined thorow with velvet, one that hath good store of coin in his purse, rings on his fingers, a watch in his pocket, which he will valew at above twenty pounds, a very broad-laced band, a stiletto by his side, a man at his heels, willing (upon small acquaintance) to intrude himself into thy company, and still desiring further to insinuate with thee; then take heed of a Jesuite of the prouder sort of priests."

In the South Kensington Museum is a metal gilt watch, the dial-plate of which is engraved with the Entombment of Christ, after Lucas van Leyden. The maker was "Nicholas Lemandre, à Blois, 1630." The length is five inches and one-eighth, and the width four inches and one-eighth. This watch was sold at the Bernal sale for 10*l.*

In the British Museum is a watch which belonged to Milton. It was made by William Bunting, in Pope's Head

Alley, Cornhill, whose name appears in the tables of the Clockmakers' Company for that period. The face is inscribed "Ioanni Miltoni, 1631;" it shows the days of the month, and the hours, and has a glass. This watch was bequeathed to the nation by Sir Charles Fellowes, who had one of the finest collections of watches in England. It is said that a poor family in Yorkshire received a box from America as part of the effects of an aged relative, whose ancestors had emigrated to that continent soon after the time of the Commonwealth. The box contained several coins and the last-mentioned watch; and the family to whom the bequest came being poor, sold the whole to a silversmith, who was also a watchmaker. The purchaser gave the full price for the coins, but refused to give more for the watch than the value of the silver case, 2*s*. 9*d*. The works, with the face on, which looked like iron, being so much tarnished, were put away in a drawer which was frequently opened. The friction after a time showed the face or dial to be of silver, with an inscription on it. This on being deciphered, after cleaning, was found to be as above given.

In 1634 was published the 'Translation of Ariosto,' by Sir John Harington, the poet, who died early in the seventeenth century. An engraving of Harington is given on the title-page, in which picture a watch is represented on a table. In Harington's 'Orlando Furioso,' 1591, the author is depicted with an article that appears to be a watch, on which is inscribed "II tempo passa."

Probably watch-seals came into use shortly after watches became generally used. At a meeting of the British Archæological Association, held on June 8th, 1864, the Rev. E. Kell, F.S.A., exhibited a watch-seal of brass, of the first half of the seventeenth century, which was found in digging in a garden in Grosvenor Square, Southampton. The face was incised with a round shield charged with the letters J. H. H., ensigned by a coronet, and flanked by laurel branches.

The English watchmakers had by the reign of Charles I. risen to such importance, that in the year 1631 they gained a charter of incorporation under the name of the Master, Wardens, and Fellowship of the Art of Clockmaking of the City of London, and by which charter all foreign clocks, watches, and alarums were forbidden to be brought into the country. David Ramsey, who had been clockmaker to James I., was, as we have before stated, appointed the first Master; and watches made by him, and by Edward East, David Bouquet, John Midnall (about 1650), Robert Grinkin (who made the watch attributed to Cromwell, in the British Museum), Benjamin Hill, and other early members of the company, are still extant. Some of their watches are round, and others oval, for the latter form continued to be occasionally employed until towards the close of the seventeenth century. The journals of the Company show that in 1635 a brass watch was of the value of forty shillings, and in the following year another watch was of the value of 4*l.* Respecting the early prices of watches little is known; but in 1643, 4*l.* were paid to redeem a watch taken from a nobleman killed in battle.

In Hollar's interesting set of plates of the 'Four Seasons,' dated 1641, the lady representing Summer has an egg-shaped watch on her left side, depending from her girdle. And Mr. Whelan had a small silver watch of the same form, and but little subsequent in date. The dial was engraved on a flat plate of silver; it had Roman numerals, with a stud at the base of each; and the steel hand terminated in a fleur-de-lis. On the plate at the back was engraved the maker's name, "Phillipe Grebay à Londres," which told that it was the work of a Frenchman. Probably the works had been renewed at a comparatively late period, for a chain took the place of the more primitive catgut. The cover was provided with a stout concavo-convex glass, set in a steel rim, and held in its place by three studs. The keyhole was protected by a crescent-formed plate moving on a pivot. At a meeting of the British Archæological Association, held on March 26th, 1856, Dr. W. V. Pettigrew exhibited an oval silver watch of about the middle of the seventeenth century, very similar to the one belonging to Mr. Whelan, and described above, and which also was exhibited to the same Association in 1855. Within the circle of Roman numerals was engraved a view with a pedlar and dog crossing a bridge; and outside the circle the plate was adorned with rich foliage in niello work. The interior of the watch was well finished, but catgut was used instead of a chain. On the back plate was inscribed the maker's name and address, "Hans Conrad Elchinger fec. Amsterdam."

At a meeting of the Archæological Institute, held on April 13th, 1855, Mr. O. Morgan exhibited a gold enamelled hunting-watch, of about 1630 or 1640. The four subjects on the front, back, and inner side of the lid and case, represented the chief incidents in the Episode of Tancredi and Clorinda, in the 'Gierusalemme Liberata' of Tasso. At a meeting of the same Institute, held on February 6th, 1863. Lord Torphichen exhibited a curious clock-watch, striking the hours, and of skilful construction, with the name of the maker, Samuel Aspinwall, engraved upon the works. It had then lately been found at Lord Torphichen's seat, Calder House, Mid Calder, with other objects of value, in an old cabinet, which had not been opened for nearly a century. The outer case of the watch was of steel, wrought in openwork studded with silver; the inner case was of silver, likewise of openwork, and among the ornamental details were an eagle, a rose, and a lily. The dial was of silver beautifully engraved, the subject being the Accusation of Susannah by the Elders. There was only an hour-hand; the hours were struck on a fine-toned bell, serving as an inner case within the pierced work. The watch measured about two-and-a-half inches in diameter, by one inch in thicknesss. There were two seals appended, one of steel, the other of jasper, engraved with the armorial bearings of the Torphichen family. The date of this watch was about 1650 or 1660. The silver pierced work of floral designs was much in vogue in the time of Charles I. It had a hair-spring and regulator, also a very fine chain, which might have been added in place of the original catgut about 1675. The name of Samuel Aspinwall is uncommon, but in 1675 Josiah Aspinwall was admitted a brother of the Clockmakers' Company.

In the South Kensington Museum is a gold enamelled watch, with the subjects of the Holy Family, after Rubens, and the Virgin and Child, after Mignard, on the exterior. The inner sides of the case contain respectively portraits of Louis XIII. and of Cardinal Richelieu. It is of French work, about 1640–50. The diameter is two inches and a quarter. It was purchased for 20*l.* At a meeting of the Society of Antiquaries of London, held on June 5th, 1862, Mr. Frederick Ouvry exhibited two watches bearing portraits and arms. These watches were the work of a Frenchman named Vernis Martin, of the time of Louis Quinze.

Lady Fitzgerald possesses a gold enamelled watch, manufactured by order of Louis XIII. as a present to our Charles I., which may rival a modern work for its smallness. It is oval, measuring about two inches by one inch and a half across the face, and is an inch in thickness. The back is chased in high relief with the figure of St. George and the Dragon. The motto of the garter surrounds the case, which is enriched with enamel colours.

A silver alarum clock-watch of circular form, which belonged to Charles I., and was usually placed by him at his bedside, is now in the possession of Mr. W. Townley Mitford. It was presented by the king to his faithful and attached servant, Mr., afterwards Sir, Thomas Herbert, on his way to execution at Whitehall, on January 30th, 1649. "It came into possession of my family," says Mr. Mitford, "by intermarriage with the Herberts, about a century ago. Since that time it has remained in our possession." Parts of the interior mechanism of the watch were modernised about fifty years ago, and the original catgut cord was replaced by a metal one. The outer case of fine perforated work, enclosing two silver bells, on which the hours and quarters are struck, remains unaltered. "Edward East, London," is engraved inside. His name is among those of the ten assistants of the Clockmakers' Company on its first incorporation in 1631; and he gave by deed of settlement, dated June 20th, 1693, the sum of 100*l.* to the Company, in trust to pay annually to five poor labouring workmen of the art or mystery of clockmaking, who were freemen of the City of London, or to the widow of each of such workmen the sum of twenty shillings. Lady Fellows has an octangular crystal-cased watch, with an engraved dial of a recumbent female figure holding an hour-glass. The maker was Edward East. Another watch by the same maker belongs to Mr. W. A. Sanford. It is of silver, in the form of a cross, and has a crystal face, the dial being engraved with the Crucifixion and angels. It has a plain outer case of the same form. He is mentioned as the king's watchmaker, living in Fleet Street, in the 'Memoirs of the two last Years of the Reign of that unparalleled Prince, of ever blessed memory, King Charles I.,' by Sir Thomas Herbert, 1702. These memoirs contain a very particular account of the various articles presented by Charles just previous to his decapitation, and also many allusions to the relic now possessed by Mr. Mitford. It appears that the King had two watches, one of gold and the other of silver, which he customarily wound up before going to bed; and they were placed before a lamp on a stool near his bedside. On the morning of his execution, as he went from his palace, he bade Sir Thomas Herbert bring him the silver clock that hung by the bedside. In the park he asked him what the time was, and then, taking the watch in his own hand, he gave it to Herbert, and told him to keep it in memory of him. This watch, which is beautifully chased and engraved, is about three inches in diameter, and about one inch and a quarter in thickness. The back and front of it are engraved in the 'Sussex Archæological Collections,' vol. iii. p. 103; and in the 'Archæological Journal' for 1850. The gold watch was confided to Herbert's care to be delivered to the Duchess of Richmond, which duty was faithfully performed. Another watch, a gold alarum, appears by a paragraph in Herbert's narrative to have been purloined by a general officer of the Praise-God-Barebones fraternity.

In Brayley and Britton's 'Description of Cheshire,' in the 'Beauties of England and Wales,' we read that at Vale Royal Abbey, which was in 1801 the residence of Thomas Cholmondeley, and since of Lord Delamere, was a watch said to have belonged to Charles I., and to have been given by him to Bishop Juxon upon the scaffold. This watch came into the Cholmondeley family by an intermarriage with the Cowpers of Overleigh, near Chester, who were related to the Juxon family.

In Horsfield's 'Sussex,' published in 1835, we read:—"In the chancel of Ashburnham Church are kept, in a glass case lined with red velvet, some relics of the unfortunate Charles I. These consist of the shirt, with ruffled wrists (on which are a few faint traces of blood) in which he was beheaded; his watch, which at the place of execution he gave to Mr. John Ashburnham. . . . . . These articles have certainly been carefully preserved. Long were they treasured by as precious relics, fit only to be gazed upon by the devotees of the Icon Basilike. At length, however, the charm was broken by Bertram Ashburnham, Esq., who, in 1743, bequeathed them to the clerk of the parish and his successors for ever, to be exhibited as great curiosities." The correctness of the legend as to the gift of the watch by the king may be doubted. It would seem that John Ashburnham was not near the king on the morning of his execution, and certainly not upon the scaffold with his royal master; the watch therefore could not have been given to him at the place of execution.

The late president of the British Archæological Association, Ralph Bernal, had a large silver watch, which was made by Richard Bowen, of London, and is said to have been given to Colonel Hammond by Charles I. whilst at Carisbrooke. It has two cases, the outer one chased and engraved with a border of flowers, and a figure of the king praying. On the back of the inner case is engraved another praying figure of a man in a gown, with Christ above, and the following legend in a scroll:—"And what I sai to you, I sai unto all, Watch." When Charles was being removed from Carisbrooke to Hurst Castle, on November 30th, 1647, he gave Mr. Worsley, who had risked his life for him, his watch, saying, "This is all my gratitude has to give." The watch is still preserved in the family. It was exhibited at a meeting of the above association when at Shorwell in 1855, at the residence of Miss Worsley, to whom it then belonged. A drawing of it was exhibited by Mr. Joseph Lionel Williams at a meeting of the British Archæological Association, held in August, 1845. It is of silver, of rather large size, and an inch in thickness; it works with catgut instead of a chain; and the face and back of the inner case are very richly engraved. It is a repeater watch, and has nineteen openings in the outer case for the escape of the sound. It has the maker's name, "Johannes Bages, Londini, fecit."

At a meeting of the Archæological Institute, held at Gloucester in July, 1860, Mr. D. I. Niblett exhibited a watch which had been given by Henrietta Maria, the queen of Charles I., to General Rudhall.

Mr. Charles Reed, of Paternoster Row, writing to 'Notes and Queries' in 1860, says, "In 1855 I received from a person who had emigrated to Australia a bracket-clock, with a request that I would accept it as a token of his gratitude for some slight service I had been able to render him. The time-piece did not appear to be of any special value, but his letter informed me that the works were constructed from the 'celebrated Van Tromp's watch.' . . . . Upon the dial-plate I find the name of 'Booth, Pontefract.' Inside the stand I have discovered the lower half of a saucer-shaped cover of shagreen, and the works, as adapted to this clock, exactly fit into this cover. The works are evidently of foreign manufacture, the mainspring is in perfect order, and the keys are attached. The watch-face was probably removed by Booth." It is said that the watch passed from Booth with "the writings" to a George Booth, who went to America, and died at Brooklyn there. We remind our readers that the Dutch admiral, Van Tromp, sailed with his fleet through the Straits of Dover in

May, 1652; and was beaten in an engagement there by the English under Blake.

In the Ashmolean Museum, Oxford, is a watch which is said to have belonged to Oliver Cromwell. Mr. J. H. Fawkes, of Farnley Hall, had a watch which is also stated to have been once owned by him. It bore the name of Jaques Cartier as the maker, and was a clock-watch, which struck the hours; the outer case was of leather, perforated, and studded with silver. In Scott's 'Antiquarian Gleanings in the North of England' this watch is said to be a repeater, by which no doubt was meant a striking-watch. The repeating movement was of later date than this specimen. At Chequers Court, Buckinghamshire, the seat of Lady Frankland Russell, is a watch which once belonged to Cromwell. Messrs. Hawkesley and Co. have an ancient silver watch, with a glass covering to the face, the maker being Young, which also is said to have belonged to the same individual. In the 'Gentleman's Magazine' for 1808 is an engraving representing three views of a watch which formerly belonged to him, and which he took out of his fob at the siege of Clonmell, and presented to the ancestor of Colonel Bagwell, whose it then was. The name of the maker, William Clay, was engraved on the work inside. The outer, or golden circle, indicating the day of the month, revolved one division every twenty-four hours; whereby the number of the day was opposed to the index hand above. This watch is now in a private room at the British Museum.

It is related of Charles II. that when he was present at the amusements in the Mall, his watchmaker, Edward East, used to attend him, as a watch was often the stake played for. Watches would seem at this period to have been curiosities, and their internal mechanism a marvel. We find gossiping Samuel Pepys, then Secretary to the Admiralty, visiting Lord Brouncker, and feasting his eyes with a sight of the movements of his lordship's watch. In his 'Diary' he thus records his visit, which was on December 22nd, 1665: "I to my Lord Brouncker's, and there spent the evening by my desire in seeing his Lordship open to pieces and make up again his watch, thereby being taught what I never knew before; and it is a thing very well worth my having seen, and am mightily pleased and satisfied with it."

In the 'Mirror' for 1833 is an engraving representing a curious heart-shaped watch-key, the original of which was given by Charles II. to the Pendrell family, as a mark of his gratitude for their having been very instrumental in his preservation. It was the size of the engraving, and made of the heart of oak; it was about three-eighths of an inch thick, was faced on each side with a plate of silver, and was surmounted by an acorn of the same metal. On one side was engraved a branching oak, with the head of Charles II.; and on the other was the following inscription:—"Quercus Car. 2d. Conservatrix 1651." The pipe of the key was of brass. This relic was in the possession of Mrs. Cope, of No.

3, Regent Street, Westminster, who was a descendant of the Pendrell family.

To the French we are indebted for the art of painting in opaque enamels, and for the application of it to the ornamentation of watch-cases. In 1630 Jean Toutin, a goldsmith of Château Surr, and a great master in painting in transparent enamels, applied himself to the use of thick colours of different tints, which should melt with fire and yet retain their lustre. He succeeded; and, as he used thin plates of gold for the foundation of his work, this style of enamel painting became available for a variety of ornamental purposes. Toutin did not keep the discovery to himself, but generously told his fellow-artists of it. Dubie, a goldsmith, who worked for the king at the Louvre, was the first who distinguished himself in this new work. After him came Morlière, a native of Orleans, who worked at Blois, and employed himself chiefly in painting rings and watch-cases; but he was excelled by his pupil, Robert Vanquer, of Blois, who produced works superior to his master, both in design and in colour. He died in 1670. Chartière, of Blois, was celebrated for his beautiful paintings of flowers; and Huand le Puisné, for figures.

In 1849 Mr. O. Morgan exhibited at a meeting of the Society of Antiquaries a watch, the enamel case of which was the work of Toutin; the subject of the outside painting being the 'Histoire d'Apian,' as an inscription near the pendant indicated. The inside of the case was painted with landscapes in enamel; the dial also was enamelled, having a subject of figures in the centre, surrounded by a white circle, on which were marked the hours. This was perhaps the first instance of an enamel dial-plate, and as the date of this watch was about 1635, we may assume that watch-dials were first enamelled about that time.

At a meeting of the Archæological Institute held on June 6th, 1862, Mr. Morgan exhibited the following remarkable series of watches exemplifying the application of enamel to the enrichment of personal ornaments. A watch, the case of which was ornamented with flowers in opaque and transparent enamels, the date being early in the seventeenth century. An exquisitely finished enamel watch-case, the work of Jean Toutin, the inventor, the date being between 1630 and 1640, and the subject, Nymphs bathing, after Polemberg. An enamelled watch, the case being finely painted by Henry Toutin, brother of the inventor, and a goldsmith and enameller at Blois, between 1630 and 1640; the subject, a series of illustrations of the story of Tancred and Clorinda, from 'Orlando Furioso.' A watch enamelled by the same artist, date 1630 to 1640, the subject being the 'Histoire d'Apian.' An enamelled watch, the case of which was beautifully ornamented with flowers raised in relief, and enriched with diamonds. The artist of this unique specimen was not known, but the movement was by D. Bouquet, who was living between 1630 and 1640. A small watch-case exquisitely painted in brilliant colours, the artist probably being either Morlière or Vanquer, between 1630 and 1650. An enamelled watch, with subjects in illustration of the birth and early life of Christ, the painting being very fine, and the whole case enriched with turquoises, the date 1630 to 1650, and the artist unknown. Two enamelled watches, the cases being exquisitely painted by Huand le Puisné, in the latter half of the seventeenth century. An enamelled watch of very fine work, the artist being I. L. Durant, who flourished in the same century, and is mentioned by Siret in his 'Dictionary of Painters.' An enamelled watch of beautiful work by an unknown artist, the date being the latter part of the seventeenth century. The chased gold case was by H. Manby, and, together with the movement, later than the enamel. Two watches, the cases of which were enamelled on copper by a French artist named Mulsund, at the end of the seventeenth or beginning of the eighteenth century. And two watch-cases of Battersea enamel, about 1750. At the same meeting the following gentlemen exhibited the under-mentioned specimens of watch enamelling:—By the Earl Amherst, an enamelled watch of the seventeenth century; on one side was represented the Holy Family, and on the other, St. Catherine; the movement bore the name of "Auguste Bretonneau, à Paris." By Mr. T. M. Whitehead, a beautiful cruciform gold watch, or *montre d'abbesse*, elaborately enamelled in opaque colours. On the face, which was protected by a crystal, was seen Christ, with the emblems of the Passion; and on the back, the Crucifixion. This was by a German artist, late in the seventeenth century, and resembled the works of Dinglinger, of Dresden. The movement bore the name of "Johannes Van Ceulen, Hagæ," and had a pendulum spring, a mechanism not known before 1675. By Mr. A. W. Franks, a small enamelled watch painted by Huand the younger, and signed "Huand le Puisné fecit." By Sir Charles Anderson, Baronet, a circular enamelled plate, probably for a watch-case: it bore the arms of James IV., Duke of Lenox, K.G., Lord Warden of the Cinque Ports, and Hereditary High Admiral of Scotland, who died in 1655. In a bordure round the achievement were introduced anchors in allusion to his naval office. This was painted in colours on a white ground. By Mr. W. Russell, an exquisitely enamelled watch, the movements of which were by Nicholas Bernard, of Paris. And by Mr. Colnaghi, an enamelled watch-case, exhibiting the portrait of George II.

At a meeting of the Society of Antiquaries, held on March 24th, 1859, Mr. Richard Frankum exhibited an oval gold watch, enamelled white, and studded with garnets. A large stone was on the back and front, which opened, and appeared originally to have concealed miniatures. The face was of gold, with delicate enamelled foliage. From the lower end hung a diminutive figure of St. George and the Dragon. A similar watch is represented in a portrait of Charles I.

The Earl of Stamford and Warrington has a small egg-shaped watch, the cases of which are cut out of jacinths; the cover is set round with rose-diamonds; it has an enamelled border; the face appears through the transparent jewel. This little gem is of the seventeenth century, and was made by David Gom, à Lyons.

In the South Kensington Museum is a watch in a rock-crystal case, the back of which is cut into facets. It has a black shagreen silver-mounted outer case. The length of the watch is one inch and five-eighths, and the width one inch and a quarter. It is of the seventeenth century, and was purchased at the Bernal sale for 11*l.* 10*s.* In the same museum is an octagonal watch in a rock-crystal case; the sides are in turquoise-blue glass, with gilt metal mounting. The height is two inches and five-eighths, and the width one inch and a quarter. The makers' names are Conrad and Reiger, German workmen. This watch was bought at the Bernal sale for 4*l.* 10*s.* Also an octagonal watch, the case of which is enriched back and front with dark-blue glass paste, with arabesque ornaments in gold and enamels. The height is one inch and three-quarters, and the width one inch and one-eighth. The maker's initials are A. B. L. This article was purchased at the Bernal sale for 5*l.* Also an oval watch in silver, with gilt metal mounts, richly engraved with allegorical figures, arabesques, and landscapes. The maker's name is "Niel Hubert, à Rouen;" and the date is the seventeenth century. The height is three inches, and the width two inches and one-eighth. This watch was bought at the Bernal sale for 5*l.* 10*s.* Also a silver watch, with pierced back and sides, and chased with flowers. The dial is of silver, engraved, the hours being enamelled in black. The maker's name is "Estienne Hubert, Rouen." The length is two inches and a quarter, and the width one inch and seven-eighths. It was purchased at the Bernal sale for 6*l.* 5*s.* Also a watch enamelled in gold, painted inside and out with flowers, and marked inside "Jacques Huon, à Paris." It is of the seventeenth century; its diameter is two inches and a quarter; and it was purchased for 30*l.* Also a gold watch, the cover of which is painted in enamel on both sides with groups of Charity and Faith. On the dial is a landscape subject. The maker's name is Nicholas Bernard, of Paris. The diameter is two inches and a quarter. This watch was purchased at the Bernal sale for 15*l.* 15*s.* Also a gold watch, the case of which is richly decorated with foliated scrollwork in relief, perforated and enamelled. The maker's name is "Claude Pascal, à la Haye." The diameter is two inches. This watch was purchased at the Bernal sale for 33*l.* And also a gold enamelled watch, on the back of the case of which are portraits of an Elector and Electress of Brandenburg in classical costume. In the centre of the dial are Diana and Endymion; and round the sides are landscapes in medallions. The makers' names are "Les deux Frères Heraut de Son, F. F. à Berlin." The diameter is one inch and five-eighths. This watch was purchased at the Bernal sale for 14*l*. 10*s*. Mr. R. J. Burnside has an enamelled watch of the seventeenth century, of French manufacture. Miss F. A. Cameron has a gold watch, enamelled with a classical medallion. It is attached to a chatelaine of similar character, with two seals and a key.

A correspondent of the 'Gentleman's Magazine' for 1810, states that an enamelled watch of early date had then lately come into his possession; it was the counterpart of the one which Mrs. Joyce Frankland holds in her hand in the engraving in Churton's 'Life of Dean Nowell,' from the picture of that lady in the hall of Brasenose College, Oxford, made about the middle of the sixteenth century. The watch in form resembled that which we now term a hunting-watch; but it was more than double the size of any watch at present in use. The outer case was not of gold, but of a metal which much resembled fine pale brass, or some mixed metal. The inner case was covered with figures exquisitely wrought in the most beautiful enamel. The maker's name was "Jeban (Jehan) Augier à Paris." A writer to 'Notes and Queries' says, "I have recently examined an ancient watch, which is said to have belonged to a character eminent in English history. The name of the maker of the watch inscribed on it is "Jehan Augier à Paris."

The reader will gather from the above that early watches were often very beautiful with their decorations of enamel. Many too were also most elaborately engraved and otherwise ornamented. Thus we find in the South Kensington Museum a watch set in a medallion of mother-of-pearl, decorated with cameo portraits and trophies in gilt metal. The whole is enriched with precious stones. The diameter of the watch is one inch, and of the medallion five inches and three quarters. It was purchased for 55*l*. Also a silver watch, the sides and back of which are engraved with flowers and scrolls. The dial is engraved with the hours, and the days of the month; the maker's name is J. Bock; and the diameter is two inches. This watch was purchased at the Bernal sale for 9*l*. 10*s*. Also a silver watch, the back and sides of which are engraved with figures of the Seasons and with flowers. The maker's name is Jean Rousseau; and the height of the article is two inches and a quarter, and the width one inch and seven-eighths. It was purchased at the Bernal sale for 3*l*. 5*s*. Also the three following watches, all of which were originally in the Bernal Collection:—An oval silver watch, richly chased with arabesques and scrollwork. The dial-plate is of gilt metal engraved; the length is two inches and three-eighths, and the width one inch and a half; and the maker's name is R. Ridgette. This watch was purchased for 4*l*. An octagonal silver parcel-gilt watch, the sides, back, and front of which are engraved. The maker's name is "C. Carncel à

Strasbourgh;" and the height is two inches, and the width one inch and an eighth. It was bought for ten guineas. A silver watch, the back of the case of which is embossed to resemble a rose. The maker's name is Benjamin Rotherodd; and the length is one inch and five-eighths, and the width one inch and a quarter. This article is of the seventeenth century, and it was purchased for 7*l*. 15*s*. In the same Museum is a watch in a crystal case in the form of a fleur-de-lis, by "G. Senez, horloger du Roi, à Rouen," 1660. The size of this article is two inches and a half by one inch and a half; and it was purchased for 12*l*. And also another watch in a crystal case, oval in shape, and facet cut. This is in size two inches by one inch; its date is the seventeenth century; and it was purchased for 8*l*. Mr. J. Heywood Hawkins has a small oval silver watch, engraved with flowers on the outer case. The maker was "Jo. Midnall in Fleet St." Another existing watch of gold has an embossed allegorical representation of the four Seasons at the back, and is inscribed, "Joseph Martineau, senior, London, No. 1142."

Watch-keys also were developed into very decorative articles, and some which have been preserved exhibit considerable taste in their designs and construction. In the South Kensington Museum is one in gilded bronze, representing a siren. It is of Italian cinque-cento work; the length is one inch and five-eighths; it was formerly in the Soulages Collection; and it was bought for 1*l*.

About the year 1650 metal chains were commonly substituted for the catgut cord before in general use.

The durability of watches when well made is very remarkable. One was produced in going order before a committee of the House of Commons to inquire into the watch trade, which was made in the year 1660; and many of ancient date which can still be kept going are now in the possession of the Clockmakers' Company.

In the 'Kingdome's Intelligencer' of February 4th–11th, 1661, is advertised as lost "a round high watch, of a reasonable size, showing the day of the month, age of the moon, and tides; upon the upper plate, Thomas Alcock fecit."

Lady Fellows has an oval silver-gilt plain watch, with pounced letters O. C., and a sword in front. On the back are the words "For God and the Commonwealth;" and on the dial-plate "A. Hooke, 1661." The initials and the motto seem to point to the time of Cromwell; but the date is subsequent to his rule, and in the reign of Charles II.

The following case, abridged from Lord Stair's 'Collection of Decisions of the Court of Session,' relates to a strange occurrence which took place in the Parliament of Scotland in 1662:—"The Lord Couper alleging that, being sitting in Parliament, and taking out his watch to see what hour it was, he gave it to my Lord Pitsligo in his hand, and that he refuses to restore it; therefore craves to be restored, and that he may have the value of it *pretio affectionis*, by his own oath. The Defender alleged, and offers to prove, that the Pursuer having put his watch in his hand, as he conceives, to see what hour it was, according to the ordinary civility, they being both sitting in Parliament, the Lord Sinclair putting forth his hand for a sight of the watch, the Defender did in the Pursuer's presence put it in his hand without the Pursuer's contradiction, which must necessarily import his consent and liberate the Defender. The Pursuer answered: the Defender having put forth his hand, signifying his desire to call for the watch, the Pursuer put the same in his hand—meaning that which is ordinary, to lend the Defender the watch to see what hour it was—which importeth the Defender's obligement to restore the same. The Defender's giving of the watch to Lord Sinclair was so subit an act, that the Pursuer could not prohibit, specially they being sitting in Parliament in the time; and, therefore, his silence cannot import a consent. The Lords (*i. e.* of Session) repelled the Defence; but would not suffer the price of the watch to be proven by the Pursuer's oath, but *prout de jure.*"

In a newspaper of 1678 appeared the following advertisement:—"Lost or stolen the 3d instant, a Gold Watch with a steel Chain, the Case studded with Gold, and lined with pink coloured Sattin, a bunch of small green taffata ribon ty'd to it. Who can give notice thereof to Mr. Richard Cooke next door to the Star Inn in the Strand near Chairingcross, shall have 40*s*. or if it be bought, there money again with content. The Watchmaker's name is Henricus Toung."

In the account of charges for work done by a goldsmith in February, 1679, we find the following:—"Item, for dresing a wach keey, 3*l*."

In the 'Autobiography and Anecdotes by William Taswell, D.D.,' 1651–1682, published by the Camden Society in 1853, we read:—"After this (1680), I never was without money as long as I stayed in the university. I bought several books, clothes, a silver-hilted sword, a gold watch, and many cups, besides a great number of bows and arrows, with which I exercised myself sometimes, and at no small price. In short, whatever my desire could fancy I had."

In the South Kensington Museum is a gold watch, the case of which is enamelled in blue with a border of flowers in coloured enamels. The dial-plate and interior of the case are enamelled; in the latter is an oval medallion with a figure of Minerva painted. The maker's name is "Pieter Wresback, Haghe, 1680." The diameter is one inch and three-eighths. This watch was purchased at the Bernal sale for 6*l*. Also a gold enamelled watch, the maker of which was Pierre Duhamel, 1680. The height is one inch and a half, and the width one inch and a quarter. This article likewise was bought at the Bernal sale for 7*l*. 15*s*. Also a carriage-watch, with a silver dial and richly-engraved and perforated gilt-metal case. The diameter is three inches and a quarter. The watch was purchased for 4*l.* Also a clock, or travelling watch, in gilt metal, with a red leather case, five inches square. The height is two inches and three quarters. The article was purchased for 8*l.* Also a silver carriage-watch, the outer case of which is embossed with the story of Alexander and Diogenes. The inner case is chased and perforated with arabesque ornaments. It is believed to be Augsburg work, about 1690; its height is four inches and a half, and its width four inches; and it was bought for 14*l.*

In the 'London Gazette' for October 21st to 24th, 1689, was the following advertisement:—"Lost the 21st Instant between the Hay Market near Charing Cross and the Rummer in Queen Street near Cheap-side, a round Gold Pendulum Watch of an indifferent small size, shewing the hours and minutes, the Pendulum went with a strait Spring, it was made by Henry Jones, Watchmaker in the Temple, the Out-Case had a Cypher pin'd on it, and the Shagreen much worn. If it comes to your hands, you are desired to bring it to the said Mr. Jones, or Mr. Snag, a goldsmith in Lumbard Street, and you shall have two Guineas Reward."

At a meeting of the British Archæological Association, held on December 10th, 1862, Mr. Baskcomb exhibited some articles which had been found in making alterations in the Manor House, Chislehurst, Kent. Among them was a silver watch, which was discovered in a small cupboard that had long since been built up in a wall in a passage. This watch was of the second half of the seventeenth century, and an inch and a quarter in diameter. The face was covered with a convex glass; and the dial, of brilliant emerald green translucid enamel, was surrounded by a circle of white enamel, on which the hours were marked in Roman numerals, and the half-hours with dots in black. The gilt hands were elegantly perforated, as was likewise the cock. On the plate was engraved the maker's name, "Roumieu à Rouen." The metal case was covered with black leather, decked with knot-work and numerous rosettes of silver piqué. Rouen was once famous for its watches, and the Hubert family seem to have been the chief makers there during the seventeenth century, Noel and the two Etiennes being its most noted members. The productions of Roumieu are less known than those of the Huberts.

At a meeting of the above-named Association, held on December 9th, 1863, Mr. Henry Godwin, F.S.A., of Speen Hill, Newbury, exhibited a handsome silver watch, of the second half of the seventeenth century, two inches and a quarter in diameter. The face was elegantly chased, the hours in Roman figures, and the minutes in Arabic numerals, being filled with black enamel. Beneath the dial was a kidney-shaped aperture, exposing a portion of the balance-wheel and hairspring; and on the margin of this opening was engraved, "Rich: Rooker: London." In the centre of the flat, solid, silver-gilt back-plate the name was repeated thus:—"Richard Rooker, London, 325." This number was also stamped on the inside of the case.

At a meeting of the Archæological Institute, held in 1847, Miss Burdett submitted for inspection a singular double-cased watch, the under side of the silver case of which was fashioned like the shell of a nautilus. The maker's name appeared in the interior, "Salomon Chesnon, Blois." The dial-plate was engraved with landscapes, figures, and foliated scrolls. From the character of its ornaments, the date of this article was assigned to the latter part of the seventeenth century. It is engraved full size in the 'Archæological Journal,' vol. v., p. 83. At a meeting of the last-named Institute, held on May 5th, 1854, Mr. J. E. Rolls exhibited a small watch made by "Salomon Chesnon, à Blois." It had no hands, the hour being indicated by an escutcheon engraved on a circular plate, which revolved within the circle showing the hours; this escutcheon was charged with the following coat:—On a cross engrailed, between four eagles displayed, five lions passant. The back of the inner case was engraved, representing a gentleman and a lady, who held a bow. Lady Fellows has a very small round gold watch, with a white enamelled dial, in an outer metal case of gold piqué on leather; the maker being Salomon Chesnon, à Blois. Diminutive watches enclosed in quaint cases, not infrequently enamelled, were chiefly made at Blois, in the Orleannois, a city once famous for its horology. In the museum of the Archæological

Institute is preserved a watch, rather smaller than the above-named nautilus example, in a ribbed silver case, of the same manufacture, and about the same date, the maker being "M. Alais, Blois." It was presented by the Rev. R. Wickham, of Twyford. At a meeting of the same Institute, held on April 7th, 1854, Mr. Charles Tucker exhibited a small oval watch, in the form of a shell, of silver enamelled, with a crystal over the face. The maker's name is "Henry Beraud fecit."

A great improvement was commenced in watchmaking, in 1658, by the invention of the spiral or pendulum spring, applied to the arbor of the balance, by which means effects were produced in its vibrations similar to the action of gravity on the pendulum of a clock. Prior to this invention, the performance of watches was very irregular, serving to give the time only approximately. The first idea of the pendulum spring is attributed to Dr. Hooke, the celebrated mathematician, who originated it in 1658; but in 1660 the invention was improved, and in 1675 it was skilfully carried out by Thomas Tompion, the celebrated watchmaker, under Dr. Hooke's superintendence. Tompion made a watch with a pendulum spring for Charles II., with this inscription:—"Robt. Hooke, invenit 1658. Thos. Tompion, fecit 1675." At a meeting of the Archæological Institute, held on April 4th, 1862, the Rev. Gregory Rhodes exhibited a silver watch, the movement of which had a regulating spring, and was believed to have been made under the direction of Dr.

Hooke. It had been preserved with the family tradition that it was presented by Charles II. to Captain Nicholas Tatterell, through whose loyalty that king was conveyed to France after his defeat at Worcester in 1651. He was rewarded with a pension, which was continued for three generations. A slab in the old churchyard at Brighton records his death in 1674. Mr. Morgan, who in his observations on watchmaking, in the 'Archæologia,' vol. xxiii., p. 93, describes Dr. Hooke's improvements, is of opinion that this watch was made not earlier than 1675, but probably towards the end of the seventeenth century. Prior, in his 'Essay on Learning,' says that Tompion, "who earned a well-deserved reputation for his admirable improvements in the art of clock and watch making, but particularly in the latter, originally was a farrier, and began his great knowledge in the equation of time by regulating the wheels of a jack to roast meat." He died in 1713. His portrait, engraved in mezzo-tinto by Smith after a painting by Kneller, which represents him in a plain coat, showing the inside of a watch, was published in 1697. Tompion was so jealous of his reputation as a watchmaker, that he would not let his name appear on any of his work which was not the best of its kind; and it is related that on the occasion of a person applying to him on the subject of a watch upon which his name fraudulently appeared, he at once broke it with a hammer, and presented another one to the person, saying, "Sir, here is a watch of my making."

A rival claim to the invention of the spring to regulate the action of the balance-wheel, which was undoubtedly devised by Dr. Hooke, and brought into use by Tompion, was made by Huygens, and also by a Frenchman; but the credit of this important discovery appears to be fairly due to our own country. John Hauteville, an ingenious mechanic, and the author of a great many curious pamphlets, who was born at Orleans in 1647, and died in 1724, claimed to be the inventor of the method of making watches with spiral springs.

Until nearly the close of the seventeenth century watches had only one hand, namely, that which pointed to the hours; but by the application of the pendulum spring, and the means thereby afforded for regulating the oscillations to the greatest nicety, watches now performed with such precision, that minute and wheel hands, which made the revolution of the dial every hour, were soon added, and so the smaller subdivisions of time were indicated. This improvement is said to have been made by Daniel Quare, a quaker, and a famous London clockmaker of that period. Many of the old watches were then altered to receive the spiral springs and other later improvements.

In 1676 Daniel Quare invented the repeating movement in watches, by which they were made to strike at pleasure. A repeating-watch, or a repeater, is one that is supplied with mechanism, by putting which in action, the wearer is enabled at any time to ascertain by the vibrating sound the hour within certain limits. This is usually effected by compressing a spring, which causes a hammer or hammers to strike on a bell the hours and quarters. And here we may introduce what Bolingbroke says in the fourth of his 'Letters on the Study of History,' 1711, in relation to a certain person:—"His reason had not the merit of common mechanism. When you press a watch or pull a clock, they answer your question with precision; for they repeat exactly the hour of the day, and tell you neither more nor less than you desire to know." One of the first of these repeating-watches was presented by Charles II. to Louis XIV. of France. A clergyman named Edward Barlow also invented a contrivance similar to that of Quare, and applied for a patent for the sole making of all pulling and repeating clocks and watches. The Clockmakers' Company petitioned James II. not to grant this patent; and in consequence thereof he appointed March 2nd, 1687, for hearing the reasons urged against it, before the Privy Council. The result was that Barlow's application was refused; and the king, after a trial of each of the respective repeating-watches made by Quare and by Barlow, gave the preference to the former, which fact was notified in the 'Gazette.' Quare afterwards made for William III. a highly-finished repeating-watch, which is still in a good state of preservation.

The identical watch that Quare made for James II., and that was preferred by him, was in the possession of John

Stanton, of Benwell, near Newcastle-upon-Tyne, when, in a letter to the 'Morning Chronicle,' dated December 11th, 1823, he gave the following description of it:—"The outer case, which is of very pure gold, is embossed with the king's head in a medallion, under which on the right is Fame in the clouds, with a trumpet at her mouth, which is held in her left hand; in her right is a wreath, which she is raising as if to crown him. On the left are two winged boys supporting the royal crown; under them a tower and fortifications, on which a flag is flying; under all is the sea running close up to a fort, and on the sea is a ship under sail. This case is also beautifully engraved and pierced with scrollwork, ornamented with cannon, mortars, shot, shells, kettle-drums, colours, and other trophies of war, and with crowns, sceptres, and other emblems of royalty. The face is of gold with black Roman letters for the hours, and figures for the minutes. In the centre is a piece of pierced work in gold upon blue steel, representing the letters J. R. R. J., combined so as to appear like an ornamental scroll, above which is the royal crown. The box is exquisitely pierced with scrollwork intermixed with birds and flowers; about the hinge is engraved a landscape, with a shepherd sitting under a tree, playing upon a pipe, with a dog at his feet, and houses, trees, &c., in the distance. On the back of the box two circular lines are drawn, between which is the following inscription:—'James II. gloria Deo in excelsis sine pretio redimi mini malâ lege ablatun bno. Regi restituitur.' Within the circle described within the inner line is engraved a figure of Justice in the clouds, reclining upon her left elbow, the hand holding the scales; in her right hand is a sceptre, with which she points to three bishops beneath her, with an altar before them. On one side of the altar is the Tower of London, with a group of 26 men carrying bags (I presume intended to represent money). On the other side is a view of the City of London in perspective, and a group of 29 men carrying similar bags, of which there are several more lying in the foreground; under all a lion and a lamb are lying together. The watch is considerably thicker than, but otherwise not much above, the common size; and every part of the engraving beautiful and distinct. It goes accurately, and is in a perfect state of preservation."

A patent was granted to Daniel Quare on August 2nd, 1695, for the invention of a portable weather-glass or barometer. He was much respected by the trade to which he belonged. At his funeral, on March 30th, 1724, most of the watchmakers in London were present. He was interred in the Quakers' burying ground at Bunhill Fields. Several artists followed in the same line as himself, particularly Tompion, Julien Le Roy, Collier, Larçay, and Thiout.

Quare's rival, Edward Barlow, whose real name was Booth, was born near Warrington, and ordained in the English College at Lisbon. He took the name of Barlow from his godfather, Ambrose Barlow, a Benedictine, who suffered at Lancaster for his religion. Dodd, in his 'Church History,' says of Barlow:—"He has often told me that at his first perusing of Euclid, that author was as easy to him as a newspaper. His name and fame are perpetuated for being the inventor of the pendulum watches; but according to the usual fate of most projectors, while others were great gainers by his ingenuity, Mr. Barlow had never been considered on that occasion, had not Mr. Thompson (accidentally becoming acquainted with the inventor's name) made him a present of 200*l*. He published a treatise on the origin of springs, wind, and the flux and reflux of the sea, 8vo, 1714, and died about two years afterwards, nearly eighty-one years of age."

The Rev. G. S. Wasey has a magnificent gold repeater by Tompion, apparently as good as ever it was, and weighing over six ounces. Madame Beauzalie, of Paris, had a large gold repeating-watch, set with diamonds, and that chimed the hours. It formerly belonged to the King of Spain. In the South Kensington Museum is a gilt-metal repeating-watch, chased and pierced, the sides being ornamented with hunting subjects, and the back with scrollwork; the dial is engraved and gilt. The maker's name is "Thomas Taylor, in Holbourn;" the diameter is two inches and a quarter; and the article was purchased at the Bernal sale for 3*l*. 10*s*.

In the 'Philosophical Transactions' appears an account by Godfrey William de Leibnitz of his portable watches, in

1675. This eminent mathematician and philosopher was born at Leipzig in 1646, and died in 1716.

In 1680 Clement, a London clockmaker, invented the anchor-escapement; and in 1695 Thomas Tompion invented the cylinder-escapement, with the horizontal wheel; but this plan was not brought into use until the following century.

The engraved portrait of Edward Backwell, Alderman of London, who died in 1679, was published; it represents him in a flowered gown, with a watch. In a gallery at Hampton Court Palace is the portrait of a man with a watch in his hand, the artist being Peter Van Aelst, who flourished in the seventeenth century.

There is extant a round silver watch, with double case, of the time of William III., which measures upwards of two inches in diameter, and one inch and a quarter in thickness. In the centre of the dial-plate is chased a shield, with an escalop-shell and a satyr's head; and on it is engraved the maker's name, "Wise, Reading." The hours are in Roman numerals, the minutes marked with strokes, with the five minutes distinguished in Arabic numbers on ovals outside the circle of strokes. The figures are engraved on silver filled in with black enamel. The hands are of blue steel. On the back-plate is inscribed "Luke Wise, Reading;" and the cock is very richly wrought.

A touch-watch is one by which the time of day may be felt. Brequet invented *une montre de touche*, in which the hours were indicated by eleven projecting studs round the rim of the case, while the pendant marked twelve o'clock. In the centre of the back of the case was placed an index or hand, which when moved forward would stop at the portion of the hour indicated by the watch, so that by means of the studs and pendant the time could be easily felt and counted. For instance, at half-past two the index would stop in the middle of the space between the second and third studs from the pendant. To the blind, as well as to persons travelling by night, these touch-watches must have been invaluable. The subjoined advertisements for the recovery of touch-watches, which had been stolen, occur in newspapers of the years 1678 and 1692 respectively:—"Lost out of a person's pocket, upon Wednesday night, being the 27th of November, betwixt Princes Street and St. John's Street (Clerkenwell), an old flat silver watch, with a studded case, and little knobs against each hour to feel what's o'clock. It was made by Mr. John Bayes. Whoever brings it to Mr. Adams, a barber, in Stanhope Street, near Clare Market, shall have twenty shillings reward." In the 'London Gazette,' No. 2844, February, 1692, we read:—"Lost, between Pickadilly and St. James Street, on Monday last, in the evening, a gold watch, made fast in a gold studded case, with high pins on each hour; made by Mr. Taylor, at the end of Fetter Lane, Holborn. Whoever brings it to Mr. Harrison, goldsmith, at the 3 Flower de Luces, in the Strand, shall have a guinea reward." The late Duke of Wellington possessed a fine watch of this description, which was presented to him by the King of Spain. A touch-watch for the use of the blind was exhibited by Mr. Dent in 1851; it was of his own manufacture.

A satirical poem, published in 1690, entitled 'Mundus Muliebris,' describing a fashionable lady's toilette, says:—

"Gold is her toothpick, gold her watch is,

And gold is everything she touches."

In the same poem reference is made to the lady's diamond croche, which was the hook to which were chained the wearer's watch, seals, intaglios, and the like.

The English watchmakers of the seventeenth century became so famous in their craft, that lest inferior articles should be sold abroad as their productions, a law was passed in 1698 obliging all makers to put their names on their watches.

Lady Fellows has the following watches, all of which are of the seventeenth century:—A silver watch, the outer case of which is in the form of a skull; on the back is an hour-glass and the words, "Incertite hora," "Æsterna respice," and other sentences. The case opens at the lower jaw. The maker was J. C. Vuolf. A silver cross opening to contain a watch, but the works are absent. The numerals are engraved, and outside is a lamb engraved; at the lower end is a seal, marked with the sacred monogram. A crystal-cased watch by Jean Rousseau, in the form of a cross, with gilt engraved dial-plate and small silver dial, with a landscape in the centre. An oval metal-gilt watch by John Wrighte, engraved with sacred subjects and inscriptions; round the side is a silver band of scriptural subjects, and inside the lids the crucifix and the offerings of the Magi. A gilt-metal oval watch, with a silver belt, inscribed "Vigilate et orate, quia nescitis horam." It has an engraved dial-plate, with a landscape in the centre, and a silver ring with numerals. A metal watch-case in the form of a book, the cover being pierced to show the hours marked on the dial. A square table-watch, with iron works and gilt dial-plate, fitting into a silver case. An oval silver tulip-shaped watch by Thomas Hande, the projecting leaves being chased in a chequered pattern. The silver-engraved dial has a landscape in the centre, and a crystal face. A large round watch-case without works; it has a bell, an engraved dial-plate, and a pierced back and front of arabesque scrolls. A round silver repeater and alarum watch, with a pierced and engraved side, of Bacchus on a cask, and flowers. It has a silver dial-plate, with gilt centre, and an outer shagreen case. A round silver repeater and alarum watch by Piquet, à Rennes, with a silver-gilt centred dial-plate. A round silver alarum repeater watch by Elias Weckerlin, à Zug, with pierced flower back, silver dial, gilt centre, and a steel key. A small round silver watch by J. B. Vuicar, à Zug, with a steel hand, and an engraved silver dial-plate. A crystal-cased watch by Aymé Noel, carved with fluted rays, having a silver dial and enclosed in a silver outer case. An oval crystal-cased watch by Croymarie, carved with cross-lines and pellets, engraved gilt face, and silver dial. A round silver watch-case, chased with vertical and curved lines and flowers between, with a silver dial. A small oval silver-gilt plain watch by Robert Grinkin, of London, with crystal face and silver dial. A large oval repeater-watch, with metal open-scroll side round the bell, silver back and dial. The name on the works is "Hierosme" "Grebauul." An oval metal repeater-watch, with pierced back, small silver dial, and landscape in the centre. A large silver repeater-watch by Jean Vallier, à Lyon, with pierced floriated ornament and white enamelled dial-plate. A round silver watch by Henry Terold, of Ipswich, with chased interlacing bands, and silver dial. A round silver watch by Louis Herve, with fluted rays and white enamelled face. An oval silver watch by Dupont, à Castres, with engraved side, of birds and squirrels; engraved dial-plate with indexes to show the hour, days of the week and month, age of the moon, and constellation; inside the cover is fitted a sundial, with box for a compass. An oval watch by G. Benard, with open gilt-metal sides, silver back and front; inside the cover is a sundial without the gnomon; the dial-plate is engraved with cherub, birds, and dogs, and has a silver ring with numerals. An oval gilt-metal watch by Papus, à Rennes, in plain cases, with gilt dial-plate, and silver ring of numerals. An oval silver watch by William North, of London, with gilt face, and two dials showing the day of the week and month, the hour, and the age of the moon. An oval metal repeater-watch by "On Cusin, à Autun," with plain cases and pierced side, and a gilt chatelaine with filigree centres. A plain oval metal watch, with black numerals and steel hour-hand, and no minutes. An oval metal watch, with engraved silver band round the side, silver dial, engraved with Time, Venus, and Cupid, and gilt border. An oval metal-gilt watch by Augustin Forfard, Sedan, with plain cases, silver ring of numerals on gilt ground, and steel hour-hand. An oval silver watch by C. Gillier, à Berne, with engraved borders and centre subjects, silver ring of numerals on dial-plate, engraved with a landscape, and chased border. A round silver repeater alarum-watch by Collet, à Rouen, with pierced floriated side, and silver dial, showing the hour and days of the month. A round silver watch by John Drake, in Fleet Street, with plain outer case, silver dial, and steel hand. A round silver repeater-watch by Chamberlaine, of Chelmsford, with pierced flowers at the back, and silver dial, showing the hour and days of the month. A round silver watch by "Thos Chamberlaine, de Chelmisforde," with silver dial. A silver-gilt octangular watch by Leonard Papon, à Gean, in plain case, with engraved dial, and no numerals. An octagonal silver-gilt watch by Josias Jolly, à Paris, with engraved border, and landscape in the centre of the face. An octangular metal-gilt watch in a plain case, with crystal face, and small dial; on the back is a pendulum which projects below the case. An octangular metal-gilt watch, the case being pierced at the back with geometrical ornaments, and in front with the Agnus Dei; the dial is inlaid with enamel flowers. An octangular crystal-cased watch by Pattru, cut in diamond facets, and engraved dial with Cupids and flowers. An octangular metal-gilt repeater-watch, with pierced back, engraved borders, and inlaid enamel dial. An octangular metal-gilt repeater watch, with pierced back, in floriated ornament, open sides round the bell, and silver dial with black Roman numerals. An octangular gilt repeater metal watch, with maker's initials G. H., pierced ornaments on the back, front, and side, and nielloed silver dial, with a smaller gilt dial in the centre. An octangular silver watch by Jere Johnson, of Exchange Alley, engraved with scrolls, crystal face cut in diamond facets, and silver dial with gilt border.

The following watches also of the seventeenth century belong to the undermentioned persons—Messrs. Farrer have a French gold watch, the case of which is painted on enamel, with the Holy Family and other sacred subjects. Also a gold watch by Theodore Girard, of Blois, the back and dial of green translucent enamel, white border, and engraved with flowers. Mr. George Field has a square gilt pedestal watch by Ferment, of London, with figures of Faith, Hope, and Charity at the angles; and another with two small medallions of busts of the electors of Germany on each side; four Cupids at the top; set all over with garnets and diamonds; resting on a base supported by four eagles, and garlands between. The height is two inches and three quarters. Mr. S. Addington has a round watch, with a painted enamel case of Venus and Adonis, in an outer case of leather ornamented with gold studs in flower patterns. Also a circular enamelled watch by Stieler, à Berlin, painted with the Rape of Europa, and portraits round the edge. Also a metal circular watch, with enamelled painting of Mars and Venus. Also an enamelled watch with a portrait of William III. on horseback at the battle of the Boyne in front. On the dial is painted St. George and the Dragon. It is in a gold perforated outer case. Also a watch of quatre foil form, with a ring of ruby enamel in the centre; round the side are stone medallions with blue enamelled frames; resting on a swivel frame of silver-gilt, chased with masks and dolphins. The Rev. J. Beck has a gold watch by Boguet, of London, in a painted enamel case of Venus and Adonis; on the dial is a portrait of a lady of the time of Charles II., and inside are landscapes. Lord Willoughby de Eresby has a gold repeater-watch by Graham, of London; the outer case being inlaid with enamel flowers. Also a gold watch by Dingley, of London; the case being ornamented with appliqué gold flowers enamelled; on the dial are inlaid enamel flowers. It has a single hour-hand, and no seconds. Also a gold watch by Pierre Lagisse, with enamelled back and dial, with painted male and female figures and landscapes round the sides and inside the case; and a coral-bead chain.

Also a gold watch with back ornamented with engine-turned steel, and a rosette in the centre of rubies and diamonds, diamond hands and border round the dial.

There is extant a very handsome repeating-watch, which is still in good order, and was made by Simon De Charmes, an eminent London watchmaker, who flourished about the beginning of the last century. He built a house at Hammersmith, which is now called Grove Hall. The estate contained about twenty-five acres. About 1730 his son, David De Charmes, resided here, and was buried in the churchyard, in which is now a gravestone to the memory of several members of this family. It is said in Faulkner's 'History of Hammersmith,' that the ancestor of this family came to England at the time of the revocation of the Edict of Nantes.

Printed in Great Britain
by Amazon

30424510R00037